Guidelines for Instructional Materials on Refuge Chamber Setup, Use, and Maintenance

Katherine A. Klein, Ph.D. and Erica E. Hall

Department of Health and Human Services
Centers for Disease Control and Prevention
National Institute for Occupational Safety and Health
Pittsburgh Research Laboratory
Pittsburgh, PA
July 2009

This document is in the public domain and may be freely copied or reprinted.

Disclaimer

The purpose of this document is to provide manufacturers and mine operators with guidance on the development of instructional materials for refuge chamber set up, use and maintenance in support of the Mine Improvement and New Emergency Response (MINER) Act of 2006. The recommendations made are not intended to serve as a legal interpretation of any regulation or standard issued by the Mine Safety and Health Administration or to substitute for manufacturer-supplied materials. Rather, this document offers suggestions to aid those responsible for instructing miners in the operation of refuge chambers. Mention of any company or product does not constitute endorsement by the National Institute for Occupational Safety and Health (NIOSH). In addition, citations to publications and websites external to NIOSH do not constitute NIOSH endorsement of the sponsoring organizations or their programs or products. Furthermore, NIOSH is not responsible for the content of these publications or websites. All web addresses referenced in this document were accessible as of the publication date.

Ordering Information

To receive documents or other information about occupational safety and health topics, contact NIOSH at

> Telephone: **1–800–CDC–INFO** (1–800–232–4636)
> TTY: 1–888–232–6348
> e-mail: cdcinfo@cdc.gov
>
> or visit the NIOSH Web site at **www.cdc.gov/niosh**.

For a monthly update on news at NIOSH, subscribe to NIOSH *eNews* by visiting **www.cdc.gov/niosh/eNews**.

DHHS (NIOSH) Publication No. 2009–XXX

July 2009

SAFER • HEALTHIER • PEOPLE™

Guidelines for Instructional Materials on Refuge Chamber Setup, Use, and Maintenance ... 1
   Introduction 5
Best Practices for Creating Effective Refuge Chamber Manuals 6
   Introduction 6
      Manual Checklist 6
   Manual Content Recommendations 7
   Quick Start Guide 7
   Gas and Atmosphere Monitoring 9
      Oxygen Supply 9
      Carbon Dioxide Scrubber 10
   Supplies/Amenities 10
      Water 10
      Food 10
      First Aid 10
      Toilet 11
      Lighting 11
   Communications 11
   Psychological and Physiological Health 11
      Psychological Health 11
      Physiological Concerns 12
      CO_2 Poisoning 12
      Oxygen Deficiency 12
      Oxygen Toxicity 12
      Movement inside the Chamber 13
   Exiting the Chamber 13
   Troubleshooting Guide 13
   Characteristics of a Good Manual 14
      Format and Style Recommendations 14
      Graphic Recommendations 15
      Other Considerations 15
Recommendations for Items to Include In a Refuge Chamber 16
   Introduction 16
   Items to Include in a Refuge Chamber 16
   First Aid 16
   Lighting 17
   Sanitation and Cleaning Supplies 18
   Items to Help Pass the Time 18
   Repair Materials and Tools 18

    Additional Supplies 19
    Placards 19
        Outside the Refuge Chamber 19
        Inside the Refuge Chamber 19
        Lithium Hydroxide Curtains—Deployment Schedule 20
        Soda Lime Cartridge Change-out Schedule 20
        Oxygen Flow Chart 20
        Photos and Graphics 20
Best Practices for Manuals for Setting Up, Inspecting, and Moving a Refuge Chamber 21
    Introduction 21
    Preparing for a Refuge Chamber 21
        Selecting a Site 21
        The Setup 21
    Maintenance and Inspections 22
        Inspections 22
        Tamper-Proof Seals 23
        Gauges 23
        Mine Phone Connections 23
        Expiration Dates 23
        Parts and Accessories 24
    Moving a Refuge Chamber 24
        Preparation 24
        Securing to Transport Vehicle 24
        Damage Assessments 24
References 25
Appendix: Task Analysis 27

Introduction

Refuge chambers may potentially save the lives of miners during a mine emergency. For this reason, it is crucial that miners know how to operate them. Unfortunately, because refuge chambers provide so many services, they can be very complicated and difficult to operate. Therefore, NIOSH has created this document with suggestions for developing manuals and educational materials.

A multidisciplinary team comprised of NIOSH engineers, sociologists, psychologists, health communication professionals, and geologists developed the recommendations contained in this document. These recommendations are based on an evaluation of manufacturers' instruction manuals from both the U.S. and globally, interviews with over 20 mining and safety experts, and an extensive literature review. This research, focused on the best practices for refuge chambers, led to the formulation of this document. It is intended to offer suggestions to manufacturers and mine operators on how to create effective and easy-to-understand training manuals for miners as well as tips to create the most comfortable and usable refuge chambers. It should be noted that these recommendations are not meant to substitute for manufacturer-supplied materials but rather to be used in conjunction with manufacturer's materials. Manufacturers should always be consulted for up-to-date information about their chamber.

Although different states and different mines refer to underground refuges by different names, this guide will simply refer to underground refuges nearest to the face as refuge chambers, whether inflatable from a skid or constructed from steel. Other popular terms for refuge chambers are rescue chambers, rescue shelters, and refuge shelters. Refuges that are nearer to the shaft, whether a prefabricated refuge chamber or one built into a crosscut, will be referred to as outby refuges because of their location in the mine. Outby refuges can be permanent, semipermanent, or portable and are usually located at every other SCSR cache. Outby refuges are sometimes called hardened rooms, outby shelters, and in-place shelters. When discussing outby refuges and refuge chambers collectively, this guide will refer to them as underground refuges. Other terms for underground refuges include refuge alternatives, locations of safety, and safe havens.

This document is designed to be read either cover to cover or by specific section. It is recommended that on first read, this guide be read cover to cover. In subsequent readings, manufacturers and mine operators may find it easier to use the table of contents page to locate specific information on parts of the refuge chamber.

Best Practices for Creating Effective Refuge Chamber Manuals

Introduction

This section is primarily intended for refuge chamber manufacturers; however, mine operators may use this section to augment the materials supplied by manufacturers.

Instruction on the operation of a refuge chamber can be complicated. Therefore, all content should be written in a clear and concise manner. Complicated concepts or procedures can often be clarified by providing an analogy or example. Also, make sure directions are detailed and accurate, such as "turn the red-colored oxygen knob counterclockwise," instead of simply stating, "turn the oxygen knob."

Also, all instructional material should be tested through a focus group of potential users who can comment on usability, readability, etc. Potential users can often make recommendations to improve the training.

Manual Checklist
In addition to providing detailed instructions, there are several topics that all manuals should address. Based on reviews of current manuals, the following have been suggested.

Table 1: Checklist of suggested items for manuals on the operation of refuge chambers

☐	When to Use
☐	Quick Start Guide
☐	Gas and Atmosphere Monitoring
☐	Oxygen Supply
☐	Carbon Dioxide Scrubber
☐	Supplies/Amenities
☐	Water
☐	Food
☐	First Aid
☐	Toilet
☐	Lighting
☐	Communications
☐	Psychological and Physiological Health
☐	Psychological Health
☐	Physiological Concerns
☐	CO2 Poisoning
☐	Oxygen Deficiency
☐	Oxygen Toxicity
☐	Movement Inside of the Chamber
☐	Exiting the Chamber

Manual Content Recommendations

In addition to the suggested elements on the operation of refuge chambers, manuals should include the following general recommendations for surviving inside refuge chambers:

- Instruct miners to keep their Self-Contained Self-Rescuers (SCSR) on until the environment inside the refuge chamber has reached breathable levels.
- Include the refuge chamber's expected duration for different quantities of people. For example, if the occupancy is doubled, explain how this will affect the breathable air supply and other factors. Also, include how the refuge chamber's duration may be extended.
- Instruct miners to conserve batteries. For example, have only one cap lamp on at a time and turn off gas detectors when not in use.
- For inflatable refuge chambers, miners should be cautious of sharp objects in their pockets as these items may puncture the walls of the refuge chamber.
- Inform miners of any items within the refuge chamber that they should not sit on, such as the scrubber.
- Include a warning that water may condense on the inside walls of the chamber and pool in low spots on the floor. This should be expected and does not indicate chamber malfunctioning.
- Include a warning that the temperature and humidity inside the refuge chamber will rise. This should be expected and does not indicate chamber malfunctioning.

Quick Start Guide

In addition to a comprehensive, detailed manual it is beneficial to develop a quick start guide that isolates the steps that are critical to operating the refuge chamber. Some instruction guides for refuge chambers are complicated and list over 20 steps. For a quick start guide, the number of steps should be grouped into small, manageable numbers. Research has found that people can only remember about seven bits of information, plus or minus two [Miller 1956]. Because using a refuge chamber is not something that a miner will do on a regular basis, it is recommended that the number of steps be reduced to much less than 20. Further, if miners are in an emergency situation where they must use the refuge chamber, having groups of seven or fewer steps would make it easier to operate successfully.

One way to shorten instructions on the activation and deployment of the refuge chamber is through the use of task analysis. Task analysis involves analyzing how a task is accomplished in order to determine the steps involved. Using task analysis, the steps necessary can be broken down into tasks, subtasks, and steps. Sample task analyses for refuge chambers can be found in the Appendix. The subtasks and steps can be made available in manuals and during training, and the tasks can be developed into quick start guides. These quick start guides can be made into stickers or placards and should be placed where miners can access them before entering the chamber, such as outside the refuge chamber or inside hard hats. Once inside the chamber, miners will have a full, detailed manual at their disposal.

We have standardized our examples to the following format:
- Deploy
- Purge
- Oxygen
- Scrubber

Note: Some chambers list in their operating procedures turning on the scrubber prior to turning on the oxygen, while others state that oxygen should come first. Most likely these operations will be performed simultaneously.

Using a standardized format creates a user-friendly environment by allowing workers who change mines to easily adapt to operating different refuge chambers; contractors will especially benefit from a standardized format. Adopting a standard terminology and keeping the order of the basic steps as similar as possible will be helpful to both miners and mine trainers.

Recommended quick start guides for many types of refuge chambers follow.

Table 2: Quick start based on Task Analysis, example 1 (see Appendix)

To Deploy Chamber:
1. **Deploy** unit
2. Seal airlock—**purge**
3. Activate **oxygen** flow
4. Hang **scrubber** curtains

Table 3: Quick start based on Task Analysis, example 2 (see Appendix)

To Deploy Chamber:
1. **Deploy** unit
2. **Purge**
3. Activate **oxygen**
4. Activate **scrubber**

Table 4 below is an example of a modified quick start guide and shows how the guide can be adapted to almost every chamber. For this model (see Task Analysis, example 3 in the Appendix) it is not possible to adhere to the preferred order (Deploy—Purge—Oxygen—Scrubber) because the release valve for the oxygen is on the exterior wall of the casing that houses the inflatable bay. Therefore, the step for turning on the oxygen must be performed first, resulting in a modified order (Oxygen—Deploy—Purge—Scrubber). Note that the standardized terminology and steps were used, only the order changed.

Table 4: Quick start based on Task Analysis, example 3 (see Appendix)

To Deploy Chamber:
1. Open access panels and valves (includes **oxygen**)
2. **Deploy** - unroll bay and inflate
3. **Purge**
4. Activate **scrubber**

Table 5: Quick start based on Task Analysis, example 4 (see Appendix)

To Deploy Chamber:
1. **Deploy** unit
2. **Purge**
3. Activate **oxygen** & make-up air
4. Hang **scrubber** curtains

Gas and Atmosphere Monitoring

Although miners may already be trained in gas detection, this information should be readily available during an emergency and written as simply and clearly as possible. Manufacturers should include in their manuals information as to which gases need to be monitored, a chart with acceptable levels of each of these gases, and suggestions on how to fix an unacceptable level of a gas with instructions on adjusting the gauges.

Make sure training discusses how both the inside and outside atmosphere is monitored. Instructions should include where the monitoring system is located and how to use these systems. Training on colorimetric gas detectors may also be included since they do not need batteries to operate and, therefore, serve as a good back-up supply.

Oxygen Supply
The task of setting the air flow should be made as simple as possible. Make sure that gauges are easy to read and easily adjusted. For example, label the gauge in terms of the number of occupants instead of the oxygen liters per minute. Instructions for setting the air flow should be as specific as possible and include directions such as "Turn the red air flow knob counterclockwise." It is also important to include settings in case the chamber is above capacity and address how this may affect the oxygen supply and refuge chamber duration.

The Mine Safety and Health Administration (MSHA) also recognizes the importance of instructions for maintaining an adequate supply of oxygen. The Supplementary Information section of the final rule states that, "instructions should include topics such as adjusting oxygen flow rates and checking for loose connections, sounds of leaking gas, damage to hoses along the length or at the fittings, and broken gauges." [73 Fed. Reg. 80672, (2008)].

If the refuge chamber is equipped with extra supplies, instructions should be included on the use and operation of each. For example, if oxygen masks are included, instructions should be available on how to properly don and adjust the mask.

Carbon Dioxide Scrubber
Instructions should be included on how to operate the carbon dioxide scrubber. Provide a chart with a change-out schedule for the soda lime cartridges, filters, or lithium hydroxide curtains. If the refuge chamber includes scrubber curtains, instruct miners to label them with the time they are hung so they will know how long each curtain is in use. Also, if the carbon dioxide scrubber operates with a motor, provide instructions on how to change out the motor in case of motor failure.

Supplies/Amenities

Water
Federal regulations state that "the Emergency Response Plan (ERP) specify a minimum of 2,000 calories of food and 2.25 quarts of potable water per person per day to sustain the maximum number of persons reasonably expected to use the refuge alternative at one time" [73 Fed. Reg. 80699, (2008)]. Make sure to include how much water is allocated for each miner per day and why it is important to ration water. If there is additional water than what is required, it is especially important to specify how much water is provided for each miner. Also, make sure to include where the water is stored in the refuge chamber and how to dispose of empty bottles.

Food
Instructions should be included on the location of all food and the allocation for each miner per day. Stress the importance of rationing food. It is also important that the ingredients are listed on the packaging in case anyone has a food allergy. One severe allergy that miners might have is a legume allergy. Legume allergies usually include peanuts, peas, and lentils. It is suggested that no peanut based food be included in the stored food caches as some people may have a severe allergic reaction while even in the vicinity of peanut products. Miners may also choose to bring their lunch pails into the refuge chamber with them.

First Aid
Make sure to list what is included in the first aid kit and where it is located. If any equipment requires instructions to operate, make sure they are included. Also, miners should be instructed that if an Emergency Medical Technician (EMT) is in the refuge chamber that he or she should be in charge of first aid supplies. It would also be beneficial to include instructions on how to care for dead and injured miners [Foster-Miller 1983].

Toilet
A manual should include instructions on how to operate the toilet and information on the any cleansing chemicals. Miners should also be supplied with instructions on how and when to dispose of the waste.

Lighting
If any lighting is included in addition to the miners' cap lamps, provide instructions on its location, how to operate it, and how long the lighting will last. For example, if there is a flashlight that can last 12 hours this should be noted so that miners can conserve power if necessary. (See page 16).

Communications

According to the federal rule, refuge chambers must include provisions for communications. The law "requires a two-way communication facility that is a part of the mine communication system, which can be used from inside the refuge alternative" [73 Fed. Reg. 80700 (2008)]. Further, MSHA recognizes that new wireless technologies are being developed and recommends that they be included in the refuge chamber as they become available. Miners should be instructed on the type of communication in the refuge chamber and how and when it to use it.

Psychological and Physiological Health

Miners should also be given a realistic preview as to what it will be like to spend four days inside of a refuge chamber. NIOSH is developing refuge chamber expectations training that provides miners with information about what to expect psychologically and physiologically while inside a refuge chamber. Manufacturers may also want to include this information in instruction manuals.

Psychological Health
It is important to remember that if miners must use a refuge chamber it is in a time of emergency. Studies have shown that most people do not panic during an emergency, but some people become stressed or anxious [Sime 1983; Hodgkinson 1990]. Including a psychological section in a training manual may help miners to reduce their psychological distress by understanding that distress is normal behavior under the conditions. Also including instructions on how miners can help others who are in psychological distress may be helpful [Foster-Miller 1983]. Further, a placard with tips for reducing stress inside of the refuge chamber can be included.

Sleep is also beneficial in times of distress. However, it is important that the miners take turns between sleeping and being alert and awake. Advise miners to make a shift schedule. This will allow all occupants to get some sleep while others monitor the conditions in the chamber. Wind-up timers may be useful to remind miners to change shifts.

Physiological Concerns
A refuge chamber is designed to provide 96 hours of breathable air. In order to be able to do this, the refuge chamber environment must be closely regulated. It is possible that a miner may become ill from overexposure to a gas if the environment is not carefully regulated. Further, overexposure to a gas may occur before entering the refuge chamber. For both reasons, a section should be added to the manuals about symptoms of and treatment for gas overexposure.

Another physiological concern for miners in refuge chambers is the heat. A large number of people in a small enclosed area is going to cause the heat and humidity to rise. Miners should be trained about what the symptoms of heat stress are and how they might be able to mitigate the effects of heat stress.

CO_2 Poisoning
Include a chart detailing the symptoms, exposure limits, and recommended treatments for carbon dioxide (CO_2) poisoning. CO_2 is an odorless gas emitted by humans as they breathe. In an enclosed area, like a refuge chamber, CO_2 can build up if it is not removed with a scrubber system. Inhaling too much CO_2, which is known as CO_2 poisoning or hypercapnia, can lead to the following [CDC 2005]:

- Headache
- Dizziness
- Restlessness
- Paresthesia (tingling)
- Dyspnea (breathing difficulty)
- Sweating
- Malaise (vague feeling of discomfort)
- Increased heart rate
- High blood pressure
- Coma
- Asphyxia
- Convulsions

Oxygen Deficiency
Include a chart detailing the symptoms, exposure limits, and recommended treatments for oxygen deficiency. When the oxygen concentration is less than 18.5% the atmosphere is said to be oxygen deficient. Oxygen is necessary for the body to perform cellular metabolism and when oxygen concentrations get too low, humans may lose consciousness or even die.

Oxygen Toxicity
Include a chart detailing the symptoms, exposure limits, and recommended treatments for oxygen toxicity. When the oxygen concentration is above 23%, the environment is said to be oxygen rich. Not only will too much oxygen cause a highly flammable environment, but extreme concentrations of oxygen can also be harmful to breathe. An environment is said to be hyperoxic if the oxygen concentration is greater than 95%. High concentrations of oxygen over a prolonged period of time can damage the central nervous system and lungs [NLM 2006].

Movement inside the Chamber

Another possible physiological concern in a refuge chamber is that miners may be uncomfortable and unable to move about for a long duration. Studies have shown that people who sit in one position for long periods of time have an increased risk of developing deep vein thrombosis (DVT). DVT occurs when a blood clot forms in a large vein, usually in the legs [CDC 2008]. In a refuge chamber, however, too much movement can also be harmful. Movement may increase carbon dioxide and reduce the amount of breathable air available.

Exiting the Chamber

Miners may make a decision to exit the chamber to try to escape the mine. If miners use a refuge chamber as a way station this would mean that they are only entering the refuge chamber temporarily. Some possible uses of refuge chamber as a way station are to rest, change SCSRs, or administer first aid. The federal law suggests that, "Refuge alternatives also can be used to facilitate escape by sustaining trapped miners until they receive communications regarding escape options or until rescuers arrive" [73 Fed. Reg. 80658, (2008)]. Regardless of the reason, miners may need instructions on exiting the refuge chamber.

Make sure to educate miners about using the refuge chamber as a way station. It is important to include considerations about the oxygen supply and how this will be affected if miners do use the refuge chamber as a way station. Because every refuge chamber operates differently, it is important to note what miners will need to do if they leave the refuge chamber and try to escape. Also include information about devices that should be turned off prior to leaving the refuge chamber and instructions on what to do if they return to the refuge chamber

When exiting the refuge chamber, miners should use the airlock to avoid contaminating the chamber and to conserve purge air. It is also suggested that miners don new, un-deployed SCSRs before exiting. Do not attempt to use a previously deployed SCSR even if it was used briefly. Once an SCSR has been used, even if only briefly, the expected life is greatly reduced. If the refuge chamber has capabilities to measure the gas concentrations outside the chamber, miners should be instructed to do so. Also, instruct miners that if they can hear rescue crews to remain calm and stay inside the chamber until the crews say it is safe to leave.

Troubleshooting Guide

A separate troubleshooting guide will be quite helpful for miners by detailing more commonplace problems and how to handle them should they arise. Inflatable refuge chambers, for example, should include instructions on what miners should do if a leak develops. A list of frequently asked questions, along with their answers, has become an increasingly popular format for troubleshooting guides.

Characteristics of a Good Manual

Special attention should be paid to crafting an aesthetically pleasing and well-organized manual. Manuals that follow a logical organizational structure and are graphically pleasing are easier to understand and use. Recommendations to create a good manual follow.

Format and Style Recommendations
Although there is some debate in the research literature about choosing a font, most studies have confirmed that a serif style font is best for printed material and that a sans serif font is best for material that is viewed on a computer screen [Wilson 2001; Graham 2002]. "Serif fonts have tiny protrusions at the tops, bottoms, or edges of the letters….Sans serif fonts have no serifs" [NIOSH 2008, p. 9]. Serif style fonts include Times New Roman, Garamond, Goudy Old Style, Palatino, and Baskerville. Sans serif fonts include Arial, Tahoma, Helvetica, and Century Gothic. The following table gives samples and summarizes these recommendations.

In manuals, a 12 point font should be adequate. Larger fonts can be used for emphasis or on placards. For the refuge chamber application, manufacturers should consider using larger fonts as they will be more readable by those who are farsighted, especially if reading glasses are not supplied or if the instructions are outside the chamber. It is important to remember that refuge chambers may be used in a time of emergency so it is essential that miners are able to read the instructions. Therefore, choosing a font is an important matter and not a matter of personal preference or style.

Table 6: Choosing fonts for instruction manuals

Medium	Type of Font	Font Size	Example Fonts
Printed materials	Serif	12	Times New Roman Garamond Goudy Old Style
Online materials	Sans Serif	12	Arial Tahoma Helvetica

Other recommendations which make manuals more readable are as follows:
- Include the model(s) for which the manual applies.
- Proofread for errors.
- Include a table of contents. An index may also be helpful for larger manuals.
- Include page numbers.
- Include headings and subheadings.

Graphic Recommendations

Because refuge chambers are a relatively new concept, graphics and photos play an important role in orienting miners to the different components of a refuge chamber. Instead of drawings, photographic images should be used whenever possible. Any photos, graphics, or signs should be laminated so that they are not affected by a humid atmosphere. Color photos or liquid crystal display (LCD) screens can be used in conjunction with text to make all instructions clear. If LCD screens are used they should be intrinsically safe. If a LCD screen is used it may be possible to show videos and presentations in conjunction with photos. Color photos which show a close-up view of all important labels, monitors, and gauges are essential. Labeling and color coding all gauges and valves will also be helpful. All photos are best printed in color, but be sure to use colors that are optimal for people with color blindness. If color printing is not used, it is important that charts and photos are still readable in grayscale printing.

Other Considerations

Studies have found that having multiple sources of information may produce faster response times [Selcon et al. 1995]. Therefore, another option for those who have trouble reading small-size fonts would be to include pictorial instructions and/or audio instructions that guide miners through the activation and use of the refuge chamber. Any sort of audio equipment that is used should be intrinsically safe. These instructions could be similar to those found on defibrillators. Audio instructions would also help those who are auditory learners as opposed to visual learners. Refuge chambers will be used in an emergency situation and miners may be stressed so having multiple forms of clear and easy-to-understand instructions will increase the chances of operating a refuge chamber correctly.

It would also be fruitful to develop training materials in addition to a manual. Individuals have different learning styles and may learn best from different media. Videos, PowerPoint slides, and Flash presentations may also be used to convey important training points about refuge chambers. Manufacturers may wish to develop and conduct training using these types of media. Although NIOSH has not developed any materials for providing instruction to miners on how to operate refuge chambers, NIOSH has developed a paper and pencil training simulation for teaching miners about the types of factors they should consider when deciding whether to use a refuge chamber. It is called "Harry's Hard Choices: Mine Refuge Chamber Training." This product is available at the following website: http://www.cdc.gov/niosh/mining/products/product160.htm. NIOSH hopes to release additional training materials this year.

Recommendations for Items to Include In a Refuge Chamber

Introduction

This section provides recommendations for items to include in a refuge chamber. These recommendations can be used by both manufacturers and mine operators.

Items to Include in a Refuge Chamber

Refuge chambers are a potentially life-saving device that miners can use in an emergency. Due to limitations imposed by the mine environment, refuge chambers cannot provide an abundance of amenities. However, several things should be considered for inclusion inside a refuge chamber to make miners more comfortable and to make the refuge chamber easier to operate.

A multidisciplinary team at NIOSH evaluated several refuge chambers, conducted interviews with miners, and reviewed the literature to develop a recommended list of items to keep inside a refuge chamber.

Table 7: Checklist of items to include in a refuge chamber

☐	First Aid Kit
☐	Lighting
☐	Sanitation and Cleaning Supplies
☐	Items to Help Pass the Time
☐	Repair Materials
☐	Additional Supplies
☐	Placards
☐	CO_2 Poisoning
☐	Lithium Hydroxide Curtain Deployment Schedule
☐	Soda Lime Cartridge Change-Out Schedule
☐	Oxygen Flow Chart
☐	Photos and Graphics

First Aid

Because miners use a refuge chamber during an emergency, they may need first aid supplies to help the injured.

Some items that are recommended for inclusion follow:
- Adhesive bandages in a variety of sizes
- Blankets
- Antibiotic ointment
- Burn ointment
- Tweezers
- Sterile gloves
- Cleansing agent and/or antibiotic towelettes
- Eye wash solution
- Critical prescription medical items such as insulin, blood pressure medication, and monitoring equipment and supplies (See note below).
- A pain reliever, aspirin and nonaspirin
- Antidiarrheal medication
- Antacid tablets
- Sugar-free cough drops

Note: It is also recommended that miners add a four-day supply of their prescribed medications to this first aid kit. It should be noted that some of these medications may have a shelf life and may need to be restocked. Also, if miners have known serious allergies they might want to include an auto injector of epinephrine, or EpiPen®. Mine operators can collect these medications from their employees and include them in their refuge chambers. It is also recommended that body bags be included in case of fatalities.

More advanced medical equipment may also be added to the refuge chamber. It is foreseeable that anyone with respiratory and/or cardiac problems might find themselves in respiratory distress after taking shelter in a refuge chamber. Medical oxygen would be useful to such miners. Units, such as the compact Oxy-Viva 3, can provide resuscitation, suction, and oxygen therapy. In addition, mine operators should consider placing an intrinsically safe automated external defibrillator (AED) unit either near or in the refuge chamber. Note: At this time, there are no AED units classified as intrinsically safe, manufacturers should consider this and, hopefully, one will become available shortly.

Lighting

MSHA recommends "lighting sufficient for persons to perform tasks. Lighting is essential to allow persons to read instructions, warnings, and gauges; operate gas monitoring detectors; and perform other activities related to the operation of the refuge alternatives" [73 Fed. Reg. 80664, 80695, (2008)]. Lighting can also provide psychological comfort and is useful for reading or playing games. Therefore, permissible lighting, such as chemical lights or flashlights, in addition to cap lamps is recommended.

A window will not only allow miners to see the outside environment, but will also be psychologically beneficial.[1]

Sanitation and Cleaning Supplies

Basic sanitation and cleaning supplies should be included because miners will be eating, drinking, and using the toilet inside the refuge chamber. The following items are recommended for inclusion:

- Personal antibacterial wipes
- Disinfecting wipes or spray
- Paper towels
- Antibacterial hand sanitizer

It is also recommended that the toilet area allow for privacy. Two toilets per chamber are also recommended, if possible [Ounanian 2007a].

Items to Help Pass the Time

In addition to the above materials, items to help pass the time will greatly improve the psychological well-being of the miners. If the refuge chamber manufacturer does not supply these materials, they should be supplied by the mine operator. The following items are recommended for inclusion:

- Decks of playing cards
- Games, such as checkers
- A large chart with easy-to-read instructions on varieties of suggested games to play
- Reading material, such as books, magazines, nondenominational Bible
- Pens, pencils, and paper
- Foldable table

Repair Materials and Tools

It is recommended that a repair kit be included in the refuge chamber. MSHA recommends that "the refuge alternative is stocked with sufficient quantities of materials and tools to repair components. Materials and tools should include metal repair materials, fiber material, adhesives, sealants, tapes, and general hardware (i.e., screws, bolts, rivets, wire, zippers, and clips)" [73 Fed. Reg. 80664, 80695, (2008)]. Instructions for how to

---

[1]The government of Western Australia has made recommendations for refuge chambers in metalliferous mines stating that "adequate lighting, ... and a small window can help occupants to cope." [DoIR 2005]

repair common problems should also be included. If the refuge chamber contains a window, a replacement window may be included. For inflatable chambers, a tent patch kit should be included.

Additional Supplies

The following additional supplies should be included inside a refuge chamber:

- Manual wind-up timers
- Reading glasses
- A low-toxicity extinguishing agent that does not produce a hazardous by-product when activated
- Provisions or storage areas for equipment (for example, place empty scrubber canisters under the seat)

Placards

Although all refuge chambers will be equipped with manuals and miners will have undergone training on how to use the refuge chamber, several placards can be hung inside and/or outside the refuge chamber to remind miners of key features.

Outside the Refuge Chamber
In order for miners to be able to deploy the refuge chamber and make it operational, there are usually levers that need to be pulled or valves to be opened. Make sure to have large arrows indicating the location of important parts of the refuge chamber necessary for operation. Deployment instructions should also be posted on the outside of the chamber.

Inside the Refuge Chamber
Placards should be hung around the refuge chamber to emphasize important operational procedures. These placards should be in color and include photos whenever possible. Also, it is important that the font is large enough to be read.

It is also important to remember that charts can be used to simplify operational procedures. Therefore, charts should be provided so that miners do not need to perform any calculations. For example, some manuals instruct miners to adjust the oxygen flow at a rate of 0.5 liters per minute per occupant. In a time of emergency, it would be easier for miners to not have to make this calculation and instead have a chart that lists those values that correspond with the number of occupants.

There are several specific placards that should be placed in the chamber including
- Symptoms of CO_2 poisoning (see page 11)
- Symptoms of O_2 deficiency and abundance (see page 11)
- Chamber contents and locations
- Tips for reducing stress
- Change-out schedule of lithium hydroxide curtains or scrubber cartridges
- Oxygen flow chart based on chamber occupancy (Make sure to include levels if the chamber is above occupancy.)
- Charts indicating the recommended concentration of each gas and dangerous levels of each gas
- Signs indicating the location of important items such as waste disposal, oxygen flow gauge, carbon dioxide scrubber, equipment, and exits

Lithium Hydroxide Curtains—Deployment Schedule
If the refuge chamber is equipped with lithium hydroxide curtains, a chart should be posted inside the refuge chamber showing the number of curtains that need to be hung for various numbers of occupants and a change-out schedule. It should also be made clear to miners where these curtains should be hung. One way to do this is to post signs in the refuge chamber that say something to the effect, "Hang Curtain 1 Here."

Soda Lime Cartridge Change-out Schedule
If the refuge chamber is equipped with soda lime cartridges or filters, a schedule should be posted to detail how often the cartridges or filters need to be changed. Also make sure that the location of spare cartridges or filters is made clear.

Oxygen Flow Chart
A chart detailing the oxygen flow rate for various numbers of occupants should be posted. This chart should also include instructions about oxygen flow if the refuge chamber is over or under capacity. In order to make this as easy as possible, color coding the charts with the cylinders may be helpful. Further, diagrams or close-up photos of the instrumentation they use—such as the oxygen flow meter—may also be posted for clarity.

Photos and Graphics
Photos and graphics should also be included to assist miners in operating the refuge chamber. Pictures of important gauges, levers, and equipment should all be included. Also a map detailing where all supplies are located would be beneficial.

Best Practices for Manuals for Setting Up, Inspecting, and Moving a Refuge Chamber

Introduction

This section is intended to assist both manufacturers and mine operators with receiving, maintaining, and moving a refuge chamber. These suggestions are based on a review of manufacturer materials, literature, and interviews with stakeholders. A multidisciplinary team at NIOSH evaluated these materials to develop these suggestions.

Preparing for a Refuge Chamber

Before a refuge chamber is delivered to a mine, instructions on how to prepare for the refuge chamber should be included. Each refuge chamber may have slightly different instructions, but some basic guidelines are offered below.

It is recommended that employees from the manufacturing company assist with the setup of the refuge chamber when it is delivered. This will ensure that mine operators know how to set up the refuge chamber and have the ability to move it as mining progresses. Detailed setup instructions for the arrival of the refuge chamber should also be included so that mine employees are able to move and set up the refuge chamber on their own at a later time. Diagrams and pictures can be used to make these instructions more clear.

Selecting a Site
For the most part, refuge chambers should be stationed within a crosscut and not in the direct path of a blast wave [Ounanian 2007b]. If avoidable, refuge stations should not be located in wet areas, near overcasts, near gob seals, near the belt drive or track entries [Ounanian 2008]. Ensure that the area in which the refuge chamber will be placed is free of debris.

An important consideration when installing a refuge chamber is size. Refuge chambers differ in length, height, and width. It is critical that the dimensions of the refuge chamber be carefully considered prior to ordering. However, with inflatable chambers it is especially important to include size requirements to ensure that enough room is allocated for the full refuge chamber size when it is deployed.

The Setup
Include instructions for which direction the refuge chamber should face when it is placed in the mine. The door should also be clearly labeled to ensure that the refuge chamber is placed in such a way that the entrance is not blocked.

Also include information about what equipment needs to be connected to the refuge chamber, including phone lines.

Any necessary preparation to the mine roof, floor, or ribs before installing the refuge chamber should be specified. If the refuge chamber should be secured to the mine floor, recommendations on how to do so should be included.

Maintenance and Inspections

Refuge chambers need to be routinely inspected. This is important because they are designed to move with advancement of the mine, and this continual movement may cause damage to the unit. Further, refuge chambers are not disposable and will, therefore, have parts that need to be replaced or fixed periodically. For these reasons, refuge chambers should be properly maintained and inspected to ensure that they are functioning properly. All inspections should be recorded and a log should be kept for when components are changed out of the refuge chamber. Guidelines and recommendations related to maintenance and inspections are listed below.

Inspections

Inspection checklists, recommendations, and requirements should be included and be tailored for each refuge chamber model. A chart is a helpful way to display which components need to be inspected and how often. An example chart is shown below. It should be noted that this is simply an example of what a chart can look like and does not provide actual data about how often inspections should be conducted. Consult the manufacturer for scheduling inspections.

Table 8: Sample schedule for inspections and replacements

	Inspections						Replace		
	Daily	Monthly	Semi-Annually	Bi-Annually	3 years	5 Years	Yearly	3 Years	5 Years
External Visual – Including Tamper Proof Seal	x								
Oxygen Cylinder		x						x	
Electrical System – Including Battery Voltage				x					
Scrubber System			x						x
Gas Cylinders				x					x
Food & Water								x	

As shown in the chart above, make sure to include how often inspections for different parts need to be conducted. Initials may also be recorded to show who performed each inspection.

Mines may also wish to use lockout/tagout labels for each piece of equipment, with a place to record the date inspected/replaced and the next date that the inspection/replacement is due. An example is the inspection sticker on our car windshield or the oil change sticker that some service stations use to record when the next oil change is due.

A similar chart can also be provided to the mine so that they may mark off when inspections have been conducted.

Instructions about what to do if a part of the refuge chamber fails an inspection should be included. Be sure to include a person to contact in the case of a failed inspection. Also include who should inspect the refuge chamber and if there is a special certification process that mine employees need to complete in order to be qualified to inspect the refuge chamber.

If mine operator-supplied equipment, such as gas detectors or strobe lights, is added to the refuge chamber, remind mine operators that this equipment should also be inspected or calibrated. This information can be added to the chart above.

Tamper-Proof Seals
If the refuge chamber contains tamper-proof seals, make sure to include instructions on how to check if the seal has been tampered with and what procedure to follow if tampering has occurred.

Gauges
Instructions on how to make sure that the gauges are correctly calibrated should be included. Also, make sure to note how often the gauges need to be calibrated.

Mine Phone Connections
Instructions on how to ensure that mine phone connections are operational should be included.

Expiration Dates
Be sure to include shelf life and expiration dates of all products in the refuge chamber. Also include contact information for the company where replacements can be ordered. Further, include instructions on how to replace products inside the refuge chamber and which products, if any, need to be replaced by a company representative. This information is important to include so mine operators can anticipate delays in replacing certain items.

Parts and Accessories
Include a basic list of parts and accessories that are tailored for that particular model of the refuge chamber and information about how to order new parts.

Moving a Refuge Chamber

Preparation
Instructions for moving a refuge chamber should be similar to those for preparing for a refuge chamber. Include any necessary preparation to the mine roof, floor, and ribs of the new location before moving the refuge chamber. Also ensure that the new location for the refuge chamber is free of debris.

Include a check list or document for mine employees to fill out every time the refuge chamber has been moved to a new location. Only persons who are task trained on how to move and setup a refuge chamber should be allowed to move a refuge chamber.

Securing to Transport Vehicle
Include specific instructions on how to move the refuge chamber. If there are special considerations for securing the refuge chamber to the transport vehicle, make sure they are noted. Also the equipment that can be used to move a refuge chamber should be noted If the unit is only meant to be towed or pushed, this should also be mentioned. Also include any instructions concerning how the refuge chamber should be angled or positioned when moving to avoid shifting its contents.

Damage Assessments
Refuge chambers need to be inspected for possible damage as a result of the move. There are certain key areas of the refuge chamber that should be checked after the move, including oxygen cylinders, piping, tamper-proof seals, deployment levers or switches, and the outside surface of the refuge chamber. Also make sure that the contents have not shifted. A checklist of places and equipment to inspect for damage should be included.

References

73 Fed. Reg 80655 [2008]. Mine Safety and Health Administration: refuge alternatives for underground coal mines; final rule. (To be codified at 30 CFR Parts 7 and 75.)

CDC [2005]. NIOSH pocket guide to chemical hazards [www.cdc.gov/niosh/npg/npgd0103.html]. Date accessed: May 2008.

CDC [2008]. Deep vein thrombosis (DVT) [www.cdc.gov/ncbddd/dvt/default.htm]. Date accessed: May 2008.

Foster-Miller [1983]. Development of guidelines for rescue chambers, Volume I. Waltham, MA: Foster-Miller, Inc. Department of the Interior contract no. JO387210 for Bureau of Mines.

Graham L [2002]. Basics of design: layout and typography for beginners. Albany, NY: Delmar.

Hodgkinson P [1990]. Ways of working with panic. Fire Prev June (23):35–38.

Miller GA [1956]. The magical number seven, plus or minus two: some limits on our capacity for processing information. Psych Rev (63):81–97.

NIOSH [2008]. NIOSH identity guide [http://inside.niosh.cdc.gov/pdfs/NIOSHIdentityGuide.pdf]. Date accessed: April 2009.

NLM [2006]. Oxygen therapy—infants [www.nlm.nih.gov/medlineplus/ency/article/007242.htm]. Date accessed: September 2008.

Ounanian D [2008]. Refuge alternatives in underground coal mines. Final report Volume III. Waltham, MA: Foster-Miller, Inc. NIOSH contract no. 200-2007-20276 for NIOSH Pittsburgh.

Ounanian D [2007a]. Refuge alternatives in underground coal mines. Phase I report. Waltham, MA: Foster-Miller, Inc. NIOSH contract no. 200-2007-20276 for NIOSH Pittsburgh.

Ounanian D [2007b]. Refuge alternatives in underground coal mines. Phase II report. Waltham, MA: Foster-Miller, Inc. NIOSH contract no. 200-2007-20276 for NIOSH Pittsburgh.

Selcon SJ, Taylor RM, McKenna FP [1995]. Integrating multiple information sources: using redundancy in the design of warnings. Ergonomics *38*(11):2362–2370

Sime JD [1983]. Affiliative behaviour during escape to building exits. J Environ Psychol *3*(1):21–41.

Wilson RF [2001]. HTML email text font readability study. Web Market Today, 97 [www.wilsonweb.com/wmt6/html-email-fonts.htm]. Date accessed: May 2008.

Appendix: Task Analysis

Table 9: Task Analysis, example 1

Tasks:	Subtasks:	Steps:
1. Deploy Unit	1. Remove safety pin.	
	2. Open door.	
	3. Pull red deployment knob.	
2. Purge	1. Enter enclosure—system automatically purges.	
3. Activate Oxygen Flow	1. Set tube fill valve to OFF position.	
	2. Set door purge valve to OFF position.	
	3. Set OXYGEN supply valve to ON position.	1. Adjust the oxygen flow valve to the number of people in the tent: Turn the black knob on the oxygen flow counter-clockwise until the black ball is set to the proper number of people according to the scale on the valve. 2. Oxygen level must be checked every 30 min—levels should be between 19.5% and 22.0%. 3. Oxygen gas monitor is in the supplies box—turn it on; also the manual gas tube monitors and chemical light sticks are in this box.
4. Hang Scrubber Curtains	1. Follow arrows to scrubber curtain boxes—follow instructions.	1. The unfolded CO_2 scrubbing curtains boxes do not absorb water and may be used as a mat for people. 2. Evenly distribute curtains throughout the enclosure—note the "Place Scrubber Here" signs.
	2. Set up a minimum of one box of curtains per person.	

Table 10: Task Analysis, example 2

Tasks:	Subtasks:	Steps:
1. Deploy Unit	1. Break safety seal.	
	2. Open door.	
2. Purge	1. Slowly open slowly valves on both compressed air bottles.	
	2. Open valves on main manifold for 10 seconds.	
	3. Monitor CO level.	1. If CO < 50ppm, then go to Task 3. 2. If CO > 50ppm, then repeat Subtask 1 and 2.
3. Activate Breathable Air (Oxygen—Mine One Air)	1. Open valves slowly on all medical-grade oxygen bottles.	
	2. Open main control valve to oxygen distribution system on the back wall of chamber.	Make sure operating pressure gauge is set to 50psi.
	3. Set flow control valve at a rate of 0.5 LPM per occupant—refer to chart.	
	4. Monitor flow rate at regular intervals to ensure correct flow rate.	

Table 10 continues on next page.

Table 10: Task Analysis, example 2 (continued)

Tasks:	Subtasks:	Steps:
4. Activate Scrubber	1. Open soda lime cartridges.	Remove base from bottom of cartridge.
	2. Place soda lime cartridge into slots located on top of the "Mine One Air" CO_2 scrubber.	Examine each cartridge separately and carefully to ensure a tight seal
	3. Level the soda lime in each cartridge with leveling tool attached to scrubber.	Essential to ensure proper operation and maximum life
	4. Monitor scrubbing system.	1. If CO_2 is > 5,000ppm, then recheck cartridges for a tight seal. 2. Replace each soda lime cartridge one at a time at the end of each 16-hour interval. Important to change out one at a time to avoid unnecessary build-up of CO_2.

Table 11: Task Analysis, example 3

Tasks:	Subtasks:	Steps:
1. Open access panels and valves (includes Oxygen)	1. Open emergency and maintenance access panels.	
	2. Open purge air, oxygen, and scrubber fan drive.	
2. Deploy—unroll bay and inflate	1. Before unrolling, make sure floor is free of debris.	
	2. Unroll and extend fully.	
	3. Inflate by pulling activation cable.	
3. Purge	1. Keep SCSR on.	
	2. Check CO level.	1. Use multi-gas detector. 2. If CO < 50ppm, then go to Task 4. 3. If CO > 50ppm, then go to Subtask 3.
	3. Open purge valve, hold on until CO < 50ppm, then go to Task 4.	

Table 11 continues on next page.

Table 11: Task Analysis, example 3 (continued)

Tasks:	Subtasks:	Steps:
4. Activate Scrubber	1. Remove wood cover—duffle bag/instructions.	
	2. Set oxygen flow meter from skid.	1. Mount oxygen flow meter to side of scrubber frame. 2. Turn knob counter-clockwise on magnetic base to "on" level. 3. Set oxygen flow meter (as per chart) to the number of persons in chamber by turning knob counter-clockwise until floating ball is in the center of the line.
	3. Open soda lime cartridges from skid.	1. Unwrap 12 cartridges. 2. Place cartridges on top of scrubber tray slots. 3. Align with outside of the scrubber box. 4. Change cartridges every 24 hours for 36 people—if less or more people consult chart for time to change.
	4. Monitor atmosphere.	1. Use multi-gas detector every 2 hours. 2. If O_2 > 22%, then lower flow rate. 3. If O_2 < 19.5%, then raise flow rate. 4. If CO_2 is > 5,000ppm, then check to see if cartridges are properly seated on the foam, make sure air motor running properly, then change cartridges. If fan fails, refer to instruction manual for change-out procedures.

Table 12: Task Analysis, example 4

Tasks:	Subtasks:	Steps:
1. Deploy Unit	1. Break safety seal.	
	2. Open door.	
2. Purge	1. Keep SCSR on.	
	2. Check CO level.	1. Open package. 2. Place fiber disc in holder. 3. Use 4-gas detector. 4. If CO disc is dark grey, black or > 200 ppm, then go to Subtask 3. If not, then go to Task 3.
	3. Open purge air valve to the 1/3 valve mark. If CO disc dark grey or black or >200 ppm, then re-purge	
	4. Check CO level.	If CO disc is dark grey, black or > 200 ppm, then repeat Subtask 3. If not, then go to Task 3.
3. Activate air (oxygen and make-up)	1. Count the number of people.	
	2. Read the O_2 chart (in green).	
	3. Set the O_2 meter according to the chart.	1. Turn the knob on the flow meter (in green). Watch the ball. 2. Align the ball with the desired mark on the scale.
	4. Read the Make-up Air chart (in blue).	
	5. Set the make-up air meter.	1. Turn the knob on the flow meter (in blue). Watch the ball. 2. Align the ball with the desired mark on the scale.

Table 12 continues on next page.

Table 12: Task Analysis, example 4 (continued)

Tasks:	Subtasks:	Steps:
4. Hang scrubber curtains	1. Locate curtains in green cans under seat or in corner of storage compartment.	
	2. Count the number of people.	
	3. Read the Scrubber curtain chart.	1. Determine correct number of scrubber boxes.
	4. Hang 2 curtains on each hanger.	1. Position curtains in center of the chamber. 2. Position them so there is 1 inch between the curtains.
	5. Set scrubber timer for 12 hours (located in 6-in square box in escapeway door).	
	6. Using permanent markers, located in scrubber box, label scrubber curtains as they are hung (0,12,36) for the hour in which they are hung.	
	7. If space is available, leave old curtains hanging.	